Teacher's Guide for Assessment

- PORTFOLIO
- PROBLEM-SOLVING THINK ALONGS
- TEACHER AND STUDENT OBSERVATION CHECKLISTS
- MANAGEMENT FORMS

Grade 4

Harcourt Brace & Company
Orlando • Atlanta • Austin • Boston • San Francisco • Chicago • Dallas • New York • Toronto • London
http://www.hbschool.com

Printed in the United States of America

ISBN 0-15-311157-7

6 7 8 9 10 082 2000

CONTENTS

Overview

Description of Assessment Components

Checklists, Rubrics, and Surveys

Interaction: A Key Factor in Instruction and Assessment

The Mathematics Portfolio

Assessing Problem Solving

Management Suggestions and Forms

Overview

The assessment program in *Math Advantage* is both comprehensive and multidimensional. It allows learners a variety of opportunities to show what they know and can do, thus providing you with ongoing information about each student's understanding of mathematics. Equally important, the assessment program involves the student in self-assessment, offering you strategies for helping your students evaluate their own growth.

The *Math Advantage* assessment program is designed around the following assessment model.

Math Advantage Assessment Model

Key: **PE**–Pupil's Edition **TE**–Teacher's Edition **TGA**–Teacher's Guide for Assessment
APK–Assessing Prior Knowledge **TCM**–Test Copying Masters
PA–Performance Assessment

Formal Assessment
- Inventory Test (TCM)
- Assessing Prior Knowledge (APK)
- Chapter Review/Test (PE/TE)
- Chapter Test (TCM)
- Study Guide and Review (PE/TE)
- Multi-Chapter Test (TCM)
- Cumulative Review (PE/TE)
- Cumulative Test (TCM)

Performance Assessment
- Performance Tasks and Rubrics (PA)
- Performance Assessment and Rubrics (PE/TE)
- Interview/Task Tests and Evaluation (PA)
- Problem-Solving Think Along Response Sheets, Scoring Guides (TGA)
- Oral Presentation Scoring Rubric (TGA)
- Project Scoring Rubric (TGA)
- Cooperative Learning Checklist (TGA)

Daily Assessment
- Mixed Review and Test Prep (PE)
- Problem of the Day (TE)
- Spiral Review (TE)
- Wrap Up & Assess (TE)

Student Self-Assessment
- Math Journal (TE)
- Group End-of-Project Checklist (TGA)
- Individual End-of-Project Checklist (TGA)
- End-of-Chapter Survey (TGA)
- Attitudes Survey (TGA)

Portfolio Assessment
- Portfolio Guide, (TGA)
- Portfolio Evaluation Form (TGA)
- Family Response Form (TGA)

Daily Assessment

Daily Assessment is *assessment embedded in daily instruction.* Students are assessed as they learn and learn as they are assessed. First you observe and evaluate your students' work on an informal basis, and then you seek confirmation of those observations through other program assessments.

Math Advantage offers the following resources to support informal assessment on a daily basis.

Daily Assessment Measures

- **Mixed Review and Test Prep** in *Pupil's Edition* at end of lesson
- **Problem of the Day,** in *Teacher's Edition* at beginning of lesson
- **Spiral Review,** in *Teacher's Edition* at beginning of lesson
- **Wrap Up & Assess,** in *Teacher's Edition* at end of lesson

Options and Suggestions

✓ **Mixed Review and Test Prep** helps students review previous lessons and practice multiple-choice testing format.

✓ **Problem of the Day** kicks off the lesson with a problem that is relevant both to lesson content and the students' world. Its purpose is to get students thinking about the lesson topic and provide insights about their ability to solve a problem related to it. Class discussion may yield clues about students' readiness to learn a concept or skill emphasized in the lesson.

✓ "Child-watching" is a natural and important part of daily assessment. As you teach the lesson, you may want to make a mental note of important observations (or devise a simple way to jot them down and keep them for future reference).

✓ You may want to use **Spiral Review,** an assessment feature in the *Teacher's Edition,* to assess skills taught in previous lessons.

✓ To assess lesson content, use **Wrap Up & Assess** in the *Teacher's Edition* at the end of each lesson. This feature includes three brief assessments.

Discuss and **Write**—to probe students' grasp of the main lesson concept

Lesson Check—a quick check on students' mastery of lesson skills

✓ Depending on what you learn from students' responses to end-of-lesson assessments, you may wish to use **Problem Solving, Reteach, Practice, or Enrichment** copying masters before starting the next lesson.

Formal Assessment

Formal Assessment in *Math Advantage* consists of a series of reviews and tests that assess how well students understand concepts, perform skills, and solve problems related to program content. Information from these measures (along with information from other kinds of assessment) is needed to evaluate student achievement and to determine grades. Moreover, analysis of results can help determine whether additional practice or reteaching is needed.

Formal Assessment in *Math Advantage* includes the following measures.

Formal Assessment Measures

- **Inventory Test,** in *Test Copying Masters*
- **Assessing Prior Knowledge,** screening worksheet for each chapter in *Assessing Prior Knowledge;* shown in *Teacher's Edition*, on Chapter Overview page
- **Chapter Review/Test,** in *Pupil's Edition* and *Teacher's Edition*
- **Chapter Test,** in *Test Copying Masters*
- **Study Guide and Review,** in *Pupil's Edition* and *Teachers Edition*
- **Multi-Chapter Test,** in *Test Copying Masters*
- **Cumulative Review,** in *Pupil's Edition* and *Teacher's Edition*
- **Cumulative Test,** in *Test Copying Masters*

Options and Suggestions

✓ All formal tests are available in two formats, multiple-choice and free-response. At times, you may choose to use the multiple-choice test because its format helps prepare students for the standardized achievement tests. At other times, you may want to use the free-response test because it gives you diagnostic information about each student—information that can help you select from the various practice or reteaching options the program offers.

✓ The **Inventory Test,** available on copying masters, is a formal assessment tool that assesses how well students have mastered the objectives from the previous grade level. Test results provide information about the kinds of review your students may need to be successful in mathematics at their new grade level. You may want to use it at the beginning of the school year or when a new student arrives.

✓ Before beginning a chapter, you may wish to give a simple screening work-sheet to find out whether your students have the skills necessary for success in that chapter.

You can find a screening worksheet for each chapter in *Assessing Prior Knowledge,* a separate assessment book. A **Review Suggestion** chart accompanies the answer key for each worksheet. It gives suggestions for building the concepts and skills that results may indicate students lack. The chart suggests specific intervention lessons or More Practice exercises that can help develop each concept or skill.

- ✓ **Chapter Review/Test** is an assessment feature in the *Pupil's Edition.* Use it at the end of a chapter to reinforce learning and determine whether there is a need for more instruction or practice. Discussion of responses can help correct lingering misconceptions before students take the Chapter Test.
- ✓ The **Chapter Test,** in *Test Copying Masters,* is available in two formats—multiple-choice, Form A, and free-response, Form B. Both test the same content. The two forms permit use of the measure as a pretest and a posttest or as two forms of the posttest.
- ✓ Students can record their answers on the test itself. However, for multiple-choice tests that have no more than 50 items, you may choose to have them use the **Answer Sheet,** similar to the "bubble form" used for standardized tests, that is located in the same book as the tests.
- ✓ **Study Guide and Review** appears in the *Pupil's Edition,* at the end of every two to four chapters. This feature presents problems that relate to content across the chapters and gives examples to help students solve them. Teacher guidance is recommended.
- ✓ The **Multi-Chapter Test,** available on copying masters, deals with the same content as the study guide. It assesses understanding of key ideas and ability to demonstrate skills emphasized in the group of chapters.
- ✓ The **Cumulative Review** is similar to the Multi-Chapter Test, but its scope broadens from the group of chapters to all the chapters up to and including the current one. Discussion of student responses can help prepare students for the Cumulative Test, which is available on copying masters.
- ✓ **The Answer Key,** in *Test Copying Masters,* provides a reduced replication of the test with answers.
- ✓ Several test record forms are available for Formal Assessment; each serves a different purpose. These forms are listed at the end of this booklet in the section titled **Management Suggestions and Forms.**
- ✓ Look for patterns in test results. They can signal a need for adjustments in instruction or assessment to meet group and individual needs.

Performance Assessment

In the past, students' grades in math were based almost solely on traditional test scores. Teachers today have come to realize that the multiple-choice format of these tests, while useful and efficient, cannot provide a complete picture of students' growth. Standardized tests may show what students know, but they are not designed to show how they *think and do things*—an essential aspect of math literacy. Performance Assessment, together with other types of assessments, can supply the missing information and balance your assessment program. Perfomance Assessments, in particular, help reveal the thinking strategies students use to work through a problem, and students usually enjoy them more than standardized tests.

Math Advantage offers the following assessment measures, scoring instruments, and teacher observation checklists for evaluating student performance.

Performance Assessment Measures

- **Quarterly Extended Performance Assessments** and **Scoring Rubrics,** in *Performance Assessment*
- **Performance Assessment,** in *Pupil's Edition* and *Teacher's Edition;* Evaluation for Items 1–4 and Rubric for Problem-Solving Items, in *Teacher's Edition* under **Using Performance Assessment**
- **Interview/Task Tests and Evaluation,** in *Performance Assessment*
- **Problem-Solving Think Alongs, Response Sheets,** and **Scoring Guides,** on pages 26–30
- **Oral Presentation Scoring Rubric,** page 10
- **Project Scoring Rubric,** page 11
- **Cooperative Learning Checklist,** page 12

Options and Suggestions

✓ Before students begin a performance task, discuss how they will be evaluated. You may choose to develop scoring rubrics with your students or use those offered in the program.

✓ If you wish, interact with students as they complete a task. A question from you that encourages reflection may be all it takes to help a puzzled student proceed. You should motivate, guide, and challenge students to produce their best work—without actually doing the work for them.

✓ The assessment program at each level includes four **Performance Assessments,** each of which has four individual tasks. These tasks can help you assess students' ability to use what they have learned to solve everyday problems. Each assessment focuses on a theme. You may wish to use this type of asssessment quarterly or at the end of grading periods to help students prepare for district or state performance tests. The four tasks along with a scoring rubic and student work samples are available in the *Performance Assessment Book.*

✓ **Performance Assessment** is a two-part performance assessment that appears in the students' book every two to four chapters. The first part assesses concepts and skills. The second part assesses problem solving. Look in your *Teacher's Edition* for an evaluation checklist, a scoring rubric, and work samples to help you evaluate students' performance.

✓ The **Interview/Task Test** is an alternate form of **Performance Assessment.** It is a "one-on-one" test that facilitates teacher-student interaction. As such, it is especially useful for assessing students who perform poorly on standard tests. This evaluation criteria for each task helps you pinpoint errors in students' thinking.

✓ The **Problem-Solving Think Along** is a self-questioning performance assessment that is designed around the problem-solving process (heuristic) used in *Math Advantage*. It is available in two forms. Either form can be used to assess performance as students work through each of the four steps in the process. Each form has its own scoring guide.

Oral Response Form—a handy interview instrument to assess student performance in problem solving. Students verbalize their thinking as they work through the process.

Written Response Form—a form that individuals or groups can use to record the process they use to solve a problem

✓ The **observation checklists and scoring rubrics,** listed below, provide a way for you to evaluate three important classroom activities. Each checklist offers criteria for evaluation.

Oral Presentation Scoring Rubric—for evaluating an individual or group presentation that may be lengthy, such as one describing a project, or brief, such as one demonstrating a way to solve a problem; also a handy self-checking guide for students to use during the planning stages of a presentation.

Project Scoring Rubric—for evaluating an individual or group project. A project is an open-ended, problem-solving activity that may involve activities such as gathering data, constructing a data table or graph, writing a report, building a model, or creating a simulation.

Cooperative Learning Checklist—for evaluating a student's behavior as he or she works in a group; also for guiding discussion of ways in which a student can become a more effective group member.

Portfolio Assessment

A portfolio is is a collection of student-selected and teacher-selected work samples that represent the individual's accomplishments and growth over a period of time. The main purpose of a portfolio in mathematics is to provide both the teacher and the student with a concise—yet comprehensive—picture of the student's progress in the subject.

Support materials for building and evaluating portfolios are listed below.

Portfolio Support Materials

- **Portfolio Guide,** on page 23
- **Portfolio Evaluation,** on page 24
- **Family Response,** on page 25

Options and Suggestions

✓ Explain to students that the purpose of their Math Portfolio is to show samples of work that demonstrate their growth in mathematics. Point out that the best sample is not necessarily the neatest paper or the one with the highest score. Discuss the kinds of selections that might best show evidence of growth in what a student knows and can do. Activities that may produce especially useful work samples are identified in the chapter.

✓ Establish a basic plan that shows how many student-selected and teacher-selected work samples will go into the portfolio during a certain period of time and when they should be selected.

✓ Ask students to list on their **Portfolio Guide** each sample they select and tell what they think it shows they have learned. This reflective activity builds self-evaluation and decision-making skills and encourages students to organize their portfolios in a thoughtful manner.

✓ Discuss portfolios at regular intervals. Compare the student's current portfolio to his or her previous one rather than to those of others. Record evidence of student growth on the **Portfolio Evaluation** sheet. You may want to list some things the student might do to improve his or her next evaluation. Attach the completed evaluation sheet to the portfolio. Use it to help make conferences with students and parents a positive experience.

✓ If students take their portfolios home to share with family members, you may wish to include **Family Response,** a home-involvement letter that requests parental review of the student's portfolio. At the bottom of the sheet is a place for a family member's comments. The Family Response sheet should be returned to school with the portfolio.

Student Self-Assessment

Research shows that self-assessment can have significant positive effects on students' learning. To achieve these effects, students must be challenged to reflect on their work and to monitor, analyze, and control their learning. Their ability to evaluate their behaviors and to monitor them grows with their experience in self-assessment.

Math Advantage offers the following self-assessment tools for your use.

Self-Assessment Resources

- **Math Journal**, ideas for journal writing in *Teacher's Edition*
- **Group End-of-Project Checklist,** on page 13
- **Individual End-of-Project Checklist,** on page 14
- **End-of-Chapter Survey,** on page 15
- **Attitudes Survey,** on page 16

Options and Suggestions

✓ The **Math Journal** is a collection of student writings that may communicate feelings, ideas, and explanations as well as responses to open-ended problems. It is an important evaluation tool in math even though it is not graded. Use the journal to gain important insights about student growth that you cannot obtain from other assessments. Look for journal icons in your *Teacher's Edition* for suggested journal-writing activities.

✓ The **Group End-of-Project Checklist,** titled "How Did Our Group Do?" is designed to assess and build group self-assessment skills. The **Individual End-of-Project Checklist,** titled "How Well Did I Work in My Group?" helps the student evaluate his or her own behavior in the group.

✓ The **End-of-Chapter Survey,** titled "How Did I Do?" leads students to reflect on what they have learned and how they learned it. Use it to help students learn more about their own capabilities and develop confidence.

✓ The **Attitudes Survey,** titled "How I Feel About Math", focuses on students' attitudes about math. Use it at regular intervals to monitor changes in individual and group attitudes.

✓ Discuss directions for completing each checklist you use and tell students that there are no "right" responses to the items. Talk over reasons for various responses.

Checklists, Rubrics, and Surveys

Types Available and Their Uses

Two types of assessment tools are offered in this section: classroom observation measures to help teachers evaluate student performance and self-assessment tools to help students evaluate their own efforts. Scoring rubrics and teacher observation checklists are available for assessing students' oral presentations, projects, and participation in cooperative learning groups. Student self-evaluation checklists give students a chance to reflect upon their work as individuals and as members of a group. They also lead students to think about what they are learning and their attitudes about math.

These checklists give you information about students' confidence, flexibility, willingness to persevere, interest, curiosity, inventiveness, and inclination to monitor and reflect on their own thinking and doing, as well as appreciation of the role of mathematics in our culture.

Student's Name ______________________ Date ____________

Oral Presentation Scoring Rubric

Check the indicators that describe the student's presentation. Use the check marks to determine the student's or group's overall score.

3 Point Score Indicators: The presentation

_______ shows evidence of extensive research/reflection.
_______ demonstrates thorough understanding of content.
_______ is exceptionally clear and effective.
_______ exhibits outstanding insight/creativity.
_______ is of high interest to the audience.

2 Point Score Indicators: The presentation

_______ shows evidence of adequate research/reflection.
_______ demonstrates acceptable understanding of content.
_______ overall is clear and effective.
_______ shows reasonable insight/creativity.
_______ is of general interest to the audience.

1 Point Score Indicators: The presentation

_______ shows evidence of limited research/reflection.
_______ demonstrates partial understanding of content.
_______ is clear in some parts but not in others.
_______ shows limited insight/creativity.
_______ is of some interest to the audience.

0 Point Score Indicators: The presentation

_______ shows little or no evidence of research/reflection.
_______ demonstrates poor understanding of content.
_______ for the most part is unclear and ineffective.
_______ does not show insight/creativity.
_______ is of little interest to the audience.

Overall score for the presentation. ____________

Comments: ____

____

____

Student's Name ________________________________ Date ______________

Project Scoring Rubric

Check the indicators that describe a student's or group's performance on a project. Use the check marks to help determine the individual's or group's overall score.

3 Point Score Indicators: The group

_______ makes outstanding use of resources.

_______ shows thorough understanding of content.

_______ demonstrates outstanding grasp of mathematics skills.

_______ displays strong decision-making/problem-solving skills.

_______ exhibits exceptional insight/creativity.

_______ communicates ideas clearly and effectively.

2 Point Score Indicators: The group

_______ makes good use of resources.

_______ shows adequate understanding of content.

_______ demonstrates good grasp of mathematics skills.

_______ displays adequate decision-making/problem-solving skills.

_______ exhibits reasonable insight/creativity.

_______ communicates most ideas clearly and effectively.

1 Point Score Indicators: The group

_______ makes limited use of resources.

_______ shows partial understanding of content.

_______ demonstrates limited grasp of mathematics skills.

_______ displays weak decision-making/problem-solving skills.

_______ exhibits limited insight/creativity.

_______ communicates some ideas clearly and effectively.

0 Point Score Indicators: The group

_______ makes little or no use of resources.

_______ fails to show understanding of content.

_______ demonstrates little or no grasp of mathematics skills.

_______ does not display decision-making/problem-solving skills.

_______ does not exhibit insight/creativity.

_______ has difficulty communicating ideas clearly and effectively.

Overall score for the project. ____________

Comments: ____

____

____

Student's Name ______________________ Date ______________

Cooperative Learning Checklist

Circle the response that best describes the student's behavior.

Never Behavior is not observable.

Sometimes Behavior is sometimes, but not always, observable when appropriate.

Always Behavior is observable throughout the activity or whenever appropriate.

The student

• is actively involved in the activity.	Never	Sometimes	Always
• shares materials with others.	Never	Sometimes	Always
• helps others in the group.	Never	Sometimes	Always
• seeks the teacher's help when all group members need help.	Never	Sometimes	Always
• fulfills his or her assigned role in the group.	Never	Sometimes	Always
• dominates the activity of the group.	Never	Sometimes	Always
• shares ideas with others.	Never	Sometimes	Always
• tolerates different views within the group about how to solve problems.	Never	Sometimes	Always

Use this checklist to discuss each student's successful cooperative learning experiences and ways in which he or she can become a more effective group member.

Project ______________________ Date ____________

Group members ______________________________

GROUP END-OF-PROJECT CHECKLIST

How Did Our Group Do?

Discuss the question. Then mark the score your group thinks it earned.

How well did our group	SCORE: Great Job	Good Job	Could Do Better
1. share ideas?	3	2	1
2. plan what to do?	3	2	1
3. carry out plans?	3	2	1
4. share the work?	3	2	1
5. solve group problems without seeking help?	3	2	1
6. make use of resources?	3	2	1
7. record information and check for accuracy?	3	2	1
8. show understanding of math ideas?	3	2	1
9. demonstrate creativity and critical thinking?	3	2	1
10. solve the project problem?	3	2	1

Write your group's answer to each question.

11. What did our group do best? ______________________________

____

____

12. How can we help our group do better? ______________________________

____

____

Name ____________________ Date __________

Project ______________________________

How Well Did I Work in My Group?

Circle **yes** if you agree. Circle **no** if you disagree.

1.	I shared my ideas with my group today.	yes	no
2.	I listened to the ideas of others in my group.	yes	no
3.	I was able to ask questions of my group.	yes	no
4.	I encouraged others in my group to share their ideas.	yes	no
5.	I was able to discuss opposite ideas with my group.	yes	no
6.	I helped my group plan and make decisions.	yes	no
7.	I did my fair share of the group's work.	yes	no
8.	I understood the problem my group worked on today.	yes	no
9.	I understood the solution to the problem my group worked on.	yes	no
10.	I can explain the problem my group worked on and its solution to others.	yes	no

Name ______________________ Date ____________

Chapter ______________________________

END-OF-CHAPTER SURVEY

How Did I Do?

Write your response.

1. I thought the lessons in this chapter were ______________________________

____.

2. The lesson I enjoyed the most was ______________________________

____.

3. Something that I still need to work on is ______________________________

____.

4. One thing that I think I did a great job on was ______________________________

____.

5. I would like to learn more about ______________________________

____.

6. Something I understand now that I did not understand before these lessons is

____.

7. I think I might use the math I learned in these lessons to ______________________________

____.

8. The amount of effort I put into these lessons was ______________________________

____.

(very little some a lot)

Name ________________________________

Date ________________________________

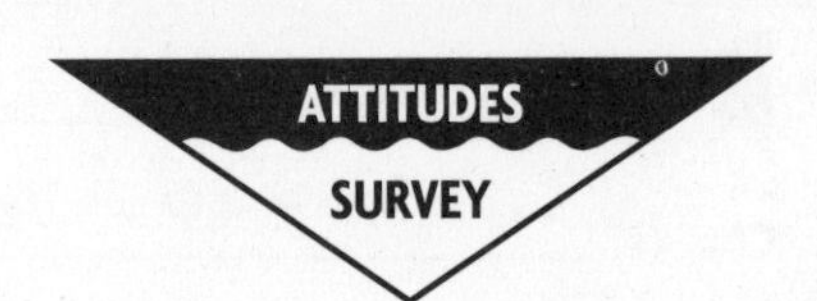

How I Feel About Math

Circle the answer that tells how you feel about each statement. If you are not sure, circle the question mark.

1. I enjoy solving problems.	Agree	?	Disagree
2. I enjoy working with others to solve problems.	Agree	?	Disagree
3. I want to complete math assignments as fast as possible.	Agree	?	Disagree
4. I think math is easy.	Agree	?	Disagree
5. I enjoy helping others in class and in groups.	Agree	?	Disagree
6. I would like extra work in math.	Agree	?	Disagree
7. I'm comfortable asking for help.	Agree	?	Disagree
8. I enjoy challenges in math.	Agree	?	Disagree
9. I like to write about how I solved a problem.	Agree	?	Disagree
10. I enjoy talking about math.	Agree	?	Disagree
11. I work very hard in math.	Agree	?	Disagree
12. I like to work alone in math.	Agree	?	Disagree
13. I like to do math outside of class.	Agree	?	Disagree
14. Math is a subject I really like.	Agree	?	Disagree

Interaction: The Key to Assessment

We believe that

- students develop mathematical power when they think consistently, communicate, and reason; draw on mathematical ideas; use tools and techniques effectively; reflect on their work; and make revisions based on that reflection.
- students learn to value diversity and to understand and appreciate the viewpoints of others when they interact with their peers.
- students take responsibility for their learning as they ask questions, formulate problems, and make decisions about what to do.

Math Advantage is a program that is centered on interaction. Students are encouraged to share ideas about what they have learned and to listen and learn what is important about the ideas being shared by others. As students work together, they agree and disagree and often learn the mathematics as a result of student-student and student-teacher interaction.

Teacher's Role

The role of the teacher in mathematics is that of a facilitator posing engaging problems and tasks to students and then challenging them to work together to solve the problems and share strategies. You can promote interaction among students in the classroom by encouraging them to explain their thinking and reasoning. Through observing and listening to students, you can guide and evaluate them in ways that are best for them.

Questioning techniques are critical strategies to ensure that the students' ideas and conclusions are based on their explorations and information. After asking a question, use **wait time** to allow the student more time to answer. If a student is unsure of an answer, restate the question or rephrase it, using different words that might help the student determine the answer. In this way, you can encourage open discussion and exploration and guide students to take an active part in their own learning.

In *Math Advantage,* students are presented with a problem or task in each lesson or project. It is important that you provide time for students to explore and to communicate with each other. You observe them working together, and based on the observations, you select questions provided that are appropriate to the students' progress on the problem or task. It is through observation and conversation that you are able to assess a student's development and depth of understanding and to guide or challenge the student to the next step to enhance his or her learning. You evaluate each student by observing and questioning as the student explains his or her thinking. The Portfolio Conference provides a time for you and the student to review the work and discuss his or her growth in mathematics.

Student's Role

Students can take active roles in their own learning. They should be encouraged to choose many kinds of manipulatives and tools when working to solve problems and tasks. Through multiple representations of the same problem, students learn to generalize and draw conclusions. They share ideas with classmates and discuss problems encountered, discoveries made, and strategies used. Through the interaction, mathematics is learned. By representing a problem in several ways, and by providing students opportunities to work on open-ended tasks and to choose their own manipulatives, you improve the quality of their interaction.

One of the most important and difficult things for a student to do is to be a good listener. Students enjoy expressing their thoughts and ideas to others, but they need to be taught to listen while others express their thoughts and ideas. Students need to understand that what others say is also important and that they might learn from them. In working together to solve problems, students should learn to question what they are doing and what they have done to test their ideas. When they share their conclusions and knowledge with others, they expand their own understanding of mathematics. By doing so, they build self-confidence and the ability to evaluate their own work.

In *Math Advantage,* students work together to construct their learning. They are encouraged to openly share and discuss ideas and strategies. Through solving problems and tasks, students develop ideas and draw conclusions about big ideas in math. They are encouraged to test their conclusions and revise them if they feel they are incorrect or if there is a better way to solve the problem.

The classroom environment should be one in which students feel comfortable sharing discoveries and ideas with others. Students should be good listeners and receptive to others' thoughts and ideas. As they talk together about mathematics, they clarify and verify their thinking and develop deeper understanding of important concepts and ideas.

Family Involvement

Family involvement in school activities helps students learn more effectively. You should keep the family informed of the content being covered as well as the growth the student is making in mathematics. Help family members understand the methods being used to evaluate the student's progress and why they are important. Family members should be encouraged to review or extend what the student is learning and experiencing at school.

Informal features designed to encourage family involvement are found throughout the program. They include the following:

- A four-page brochure that introduces family members to the *Math Advantage* program at the beginning of the school year.
- A letter describing what students will learn and suggesting activities that can be done at home to support learning is located at the beginning of each unit.
- Home Notes are located on each student activity page to keep the family informed about the mathematics concepts children are learning.
- Student Journals containing a written record of students' work and thoughts about their work can be sent home periodically to be shared with family members.
- An Assessment Portfolio with samples of the student's work, the results of the Performance Assessment Tasks, and the written assessment can be sent home periodically. In this way, the teacher presents family members with a clear picture of their child's progress. Family members should be encouraged to respond by writing in the portfolio to express the growth they see in their child's performance.

Family Involvement Tips

- Schedule periodic conferences with family members and the student to share and discuss the contents of the student's portfolio and to discuss goals for his or her mathematical growth.
- Invite family members to visit the classroom to observe or participate while the students are working on a problem or task.
- Invite students and their family members to participate in a "Family Math Night," during which they work together, using a variety of manipulatives to solve problems or tasks similar to those worked on during class time.
- Encourage family members to collect or make materials needed for math class.
- Encourage family members to help the student create a game or puzzle that uses math skills.

The Mathematics Portfolio

The portfolio is a collection of each student's work gathered over an extended period of time. A portfolio illustrates the growth, talents, achievements, and reflections of the mathematics learner and provides a means for the teacher to assess the student's performance.

An effective portfolio will

- include items collected over the entire school year.
- show the "big picture"—providing a broad understanding of a student's mathematics language and feelings through words, diagrams, checklists, and so on.
- give students opportunities to collaborate with their peers.
- give students a chance to experience success, to develop pride in their work, and to develop positive attitudes toward mathematics.

Building a Portfolio

There are many opportunities to collect students' work throughout the year as you use *Math Advantage.* A list of suggested portfolio items for each chapter is given in the Chapter Overview in your Teacher's Edition. Give students the opportunity to select some work samples to be included in the portfolio.

To begin:

- Provide a file folder for each student with the student's name clearly marked on the tab or folder.
- Explain to students that throughout the year they will save some of their work in the folder. Sometimes it will be their individual work; sometimes it will be group reports and projects, or completed checklists.
- Assign a fun activity to the entire class that can be placed into the portfolio by all students.
- Comment positively on the maps and reinforce the process. Have students place their maps into their portfolios.

Evaluating a Portfolio

The ultimate purpose of assessment is to enable students to evaluate themselves. Portfolios have the potential to create authentic portraits of what students learn and offer an alternative means for documenting growth, change, and risk-taking in mathematics learning. Evaluating their own growth in mathematics will be a new experience for most students. The following points made with regular portfolio evaluation will encourage growth in self-evaluation.

- Discuss the contents of the portfolio with each student as you examine it at regular intervals during the school year.
- Examine each portfolio on the basis of the growth the student has made, rather than in comparison with other portfolios.
- Ask the student questions as you examine the portfolio.
- Note statements that express how the student feels.
- Point out the strengths and weaknesses in the work.
- Encourage and reward students by emphasizing the growth, the original thinking, and the completion of tasks you see.
- Reinforce and adjust instruction of the broad goals you want to accomplish as you evaluate the portfolios.

What to Look For

Growth in mathematics is shown by:

- the non-standard responses that students make.
- the ways students solve a problem or attempt to solve a problem. Note the strategies they used and the reasons they succeeded or became confused.
- the ways students communicate their understanding of math problems. Do they use words, pictures, or abstract algorithms?
- unique solutions or ways of thinking.

Placing comments such as the following on students' work can have instructional benefits.

"Interesting approach."
"Is this the only solution to the problem?"
"Can you think of a related problem?"
"What about . . . ?"

Sharing the Portfolios

- Examine the portfolio with family members to share concrete examples of the work the student is doing. Emphasize the growth you see as well as the expectations you have.
- Examine portfolios with your students to emphasize their experiences of success and to develop pride in their work and positive attitudes toward mathematics.
- Examine portfolios with your supervisor to share the growth your students have made and to show the ways you have developed the curriculum objectives.

The Benefits of Mathematics Portfolios

Portfolios can be the basis for informed change in mathematics classrooms because they:

- send positive messages to students about successful processes, rather than end results.
- give you better insights as to how students understand and work problems.
- focus on monitoring the development of reasoning skills.
- help students become responsible for their own learning.
- promote teacher-student dialogue.
- focus on the student, rather than on the assignment.
- focus on the development of conceptual understandings, rather than applications of skills and procedures.

Name ______________________________

Date ______________________________

PORTFOLIO GUIDE

A Guide to My Math Portfolio

What Is in My Portfolio?	What I Learned.
1.	
2.	
3.	
4.	
5.	

I organized my portfolio this way because ______________________________

____

______________________________.

Name ______________________________

Date ______________________________

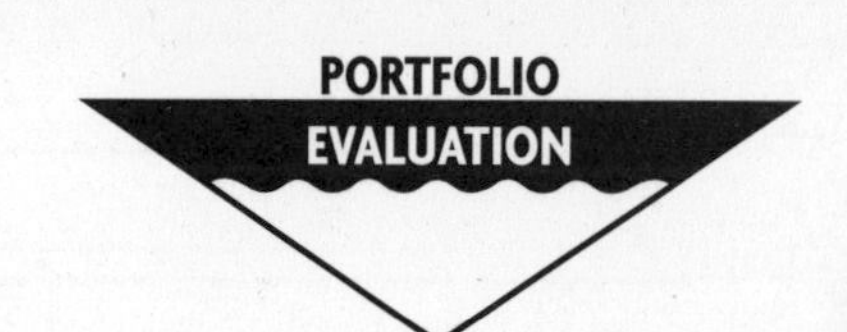

Evaluating Performance	Evidence and Comments
1. What mathematical understandings are demonstrated?	
2. What skills are demonstrated?	
3. What approaches to problem solving and critical thinking are evident?	
4. What work habits and attitudes are demonstrated?	

Summary of Portfolio Assessment

For This Review			Since Last Review		
Excellent	Good	Fair	Improving	About the Same	Not as Good

Date ______________________

Dear Family,

This is your son/daughter's math portfolio. It contains work samples that your child and I have selected to show how his or her abilities in math have grown. Your son/daughter can explain what each sample shows.

Please look over the portfolio with your child and write a few comments in the blank space at the bottom of this sheet about what you have seen. Your child has been asked to bring the portfolio with your comments included back to school.

Thank you for helping your child evaluate his/her portfolio and take pride in the work he or she has done. Your interest and support is important to your child's success in school.

Sincerely,

____

(Teacher)

Response to Portfolio:

____

(Family member)

Assessing Problem Solving

Assessing a student's ability to solve problems in *Math Advantage* involves more than checking the student's answer. It involves looking at how students process information and how they work at solving problems. The heuristic used in the program's Problem-Solving Think Along—Understand, Plan, Solve, Look Back—guides the student's thinking process and provides a structure within which the student can work toward a solution. Evaluating the student's progress through the parts of the heuristic can help you assess the student's areas of strength and weakness in solving problems.

These instruments may be used to assess students' problem-solving abilities.

Problem-Solving Think Along

Oral Response Form (See page 27.)

This form can be used by a student or group as a self-questioning instrument or as a guide for working through a problem. It can also be an interview instrument the teacher can use to assess students' problem-solving skills and can be used with the Interview/Task Test.

Problem-Solving Think Along

Scoring Guide for Oral Responses (See page 28.)

This analytic Scoring Guide, which has a criterion score for the Understand and Plan sections and one for the Solve and Look Back sections, may be used to evaluate the oral presentation of an individual or cooperative learning group.

Problem-Solving Think Along

Written Response Form (See page 29.)

This form provides a recording sheet for a student or group to record their responses as they work through each section of the heuristic. This Written Response Form is also used to record students' responses to the problem-solving items in the Interview/Task Test for each chapter.

Problem-Solving Think Along

Scoring Guide for Written Responses (See page 30.)

This analytic Scoring Guide, which gives a criterion score for the Understand and Plan sections and one for the Solve and Look Back sections, will help you pinpoint the parts of the problem-solving process in which your students need more instruction.

Name ____________________

Date ____________________

ORAL RESPONSE FORM

Problem-Solving Think Along: Oral Response Form

Solving problems is a thinking process. Asking yourself questions as you work through the steps in solving a problem can help guide your thinking. These questions will help you understand the problem, plan how to solve it, solve it, and then look back and check your solution. These questions will also help you think about other ways to solve the problem.

Understand

1. What is the problem about?

2. What is the question?

3. What information is given in the problem?

Plan

4. What problem-solving strategies might I try to help me solve the problem?

5. About what do I think my answer will be?

Solve

6. How can I solve the problem?

7. How can I state my answer in a complete sentence?

Look Back

8. How do I know whether my answer is reasonable?

9. How else might I have solved this problem?

Name ____________________

Date ____________________

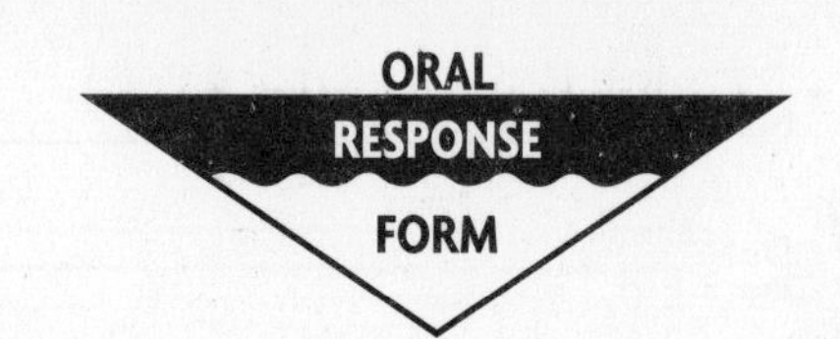

Problem-Solving Think Along:
Scoring Guide • Oral Responses

Understand
Criterion Score 4/6 *Pupil Score* ______

______ **1.** *Restate the problem in his or her own words.*
0 points - No restatement given.
1 point - Incomplete problem restatement.
2 points - Complete problem restatement given.

______ **2.** *Identify the question.*
0 points - No restatement of the question given.
1 point - Incomplete or incorrect restatement of the question given.
2 points - Complete restatement of the question given.

______ **3.** *State list of information needed to solve the problem.*
0 points - No list given.
1 point - Incomplete list given.
2 points - Complete list given.

Plan
Criterion Score 3/4 *Pupil Score* ______

______ **1.** *State one or more strategies that might help solve the problem.*
0 points - No strategies given.
1 point - One or more strategies given but are poor choices.
2 points - One or more useful strategies given.

______ **2.** *State reasonable estimated answer.*
0 points - No estimated answer given.
1 point - Unreasonable estimate given.
2 points - Reasonable estimate given.

Solve
Criterion Score 3/4 *Pupil Score* ______

______ **1.** *Describe a solution method that correctly represents the information in the problem.*
0 points - No solution method given.
1 point - Incorrect solution method given.
2 points - Correct solution method given.

______ **2.** *State correct answer in complete sentence.*
0 points - No sentence given.
1 point - Sentence given does not answer the question correctly.
2 points - Complete sentence given, answer to question is correct.

Look Back
Criterion Score 3/4 *Pupil Score* ______

______ **1.** *State sentence explaining why the answer is reasonable.*
0 points - No solution method given.
1 point - Sentence given with incomplete or incorrect reason.
2 points - Complete and correct explanation given.

______ **2.** *Describe another strategy that could have been used to solve the problem.*
0 points - No other strategy described.
1 point - Another strategy described, but strategy is a poor choice.
2 points - Another useful strategy described.

TOTAL 13/18 *Pupil Score* ______

Name ____________________

PROBLEM-SOLVING THINK ALONG

Problem Solving

Understand

1. Retell the problem in your own words. ____________________

2. List the information given. ____________________

3. Restate the question as a fill-in-the-blank sentence. ____________________

Plan

4. List one or more problem-solving strategies that you can use. ____________________

5. Predict what your answer will be. ____________________

Solve

6. Show how you solved the problem. ____________________

7. Write your answer in a complete sentence. ____________________

Look Back

8. Tell how you know your answer is reasonable. ____________________

9. Describe another way you could have solved the problem. ____________________

Problem-Solving Think Along

Scoring Guide • Written Responses

Understand	*Criterion Score* 4/6
Indicator 1: Student restates the problem in his or her own words.	***Scoring:*** 0 points - No restatement written. 1 point - Incomplete problem restatement written. 2 points - Complete problem restatement written.
Indicator 2: Student restates the question as a fill-in-the-blank statement.	0 points - No restatement written. 1 point - Incorrect or incomplete restatement. 2 points - Correct restatement of the question.
Indicator 3: Student writes a complete list of the information needed to solve the problem.	0 points - No list made. 1 point - Incomplete list made. 2 points - Complete list made.
Plan	*Criterion Score* 3/4
Indicator 1: Student lists one or more problem-solving strategies that might be helpful in solving the problem.	***Scoring:*** 0 points - No strategies listed. 1 point - One or more strategies listed, but strategies are poor choices. 2 points - One or more useful strategies listed.
Indicator 2: Student gives a reasonable estimated answer.	0 points - No estimated answer given. 1 point - Unreasonable estimate given. 2 points - Reasonable estimate given.
Solve	*Criterion Score* 3/4
Indicator 1: Student shows a solution method that correctly represents the information in the problem.	***Scoring:*** 0 points - No solution method written. 1 point - Incorrect solution method written. 2 points - Correct solution method written.
Indicator 2: Student writes a complete sentence giving the correct answer.	0 points - No sentence written. 1 point - Sentence has an incorrect numerical answer or does not answer the question. 2 points - Sentence has correct answer and completely answers the question.
Look Back	*Criterion Score* 3/4
Indicator 1: Student writes a sentence explaining why the answer is reasonable.	***Scoring:*** 0 points - No sentence written. 1 point - Gives an incomplete or incorrect reason. 2 points - Gives a complete and correct explanation .
Indicator 2: Student describes another strategy that could have been used to solve the problem.	0 points - No other strategy described. 1 point - Another strategy described, but it is a poor choice. 2 points - Another useful strategy described.

TOTAL 13/18

Management Suggestions and Forms

Description of Materials and Their Locations

- **Test Answer Sheet,** in *Test Copying Masters*
  This copying master is an individual recording sheet for up to 50 items on the multiple-choice (standardized) test format.

- **25 Test-Taking Tips,** pages 32–33
  Help your students do their best on standardized-format tests by talking over these these "do's" and "don'ts" on preparing for and taking a test.

- **Grading Made Easy,** in *Test Copying Masters*
  This percent converter can be used for all quizzes and tests. The percents given are based on all problems having equal value. Percents are rounded to the nearest whole percent, giving the benefit of 0.5 percent.

- **Individual Record Form,** in *Test Copying Masters*
  This form provides a place to enter a single student's scores on all formal tests and to indicate the objectives he or she has met. Criterion scores for each learning goal are given. A list of review options is also included. The options include activities in the *Pupil's Edition, Teacher's Edition, and Workbooks* that you can assign to the student who is in need of additional practice. One copying master is provided for each content cluster of chapters.

- **Formal Assessment Class Record Form,** in *Test Copying Masters*
  This form makes it possible to record the test scores of an entire class on a single form. Criterion scores are listed for each of the tests.

- **Performance Assessment Class Record Form,** in *Test Copying Masters*
  Use this record form to check the skills your students demonstrate on the Interview Task/Tests and on other performance assessments.

25 Test-Taking Tips

Prepare for the Test

1. Get a good night's sleep and eat a good breakfast before taking the test.
2. Tell yourself that you will do the very best that you can during the test.

Be Sure That You Understand the Directions

3. When your teacher gives special directions for the test, pay attention only to the teacher.
4. When there are directions printed on the test, read them and give each one time to sink in.
5. Reread the directions if you don't understand them.
6. Be sure that you can answer these questions.

   A. Can I write on the test itself?
   B. How many answers should I mark for each item?
   C. How many items are on the test?
   D. How long do I have to work on the test?

7. Before taking a timed test, you may want to ask your teacher to write the number of minutes remaining on the chalkboard several times while you are working.
8. Follow all test directions carefully—do *exactly* what they tell you to do.
9. If something does not seem right as you answer questions, go back and read the directions again.

Read Each Question Carefully

10. Read each question slowly. Stop and think about what is asked. If you are not sure, read the question again. If you still are not sure, go on to the next question.
11. Be on the lookout for small but important words like *only, always, all,* and *never.* Remember that these words mean *with **no** exceptions.*
12. Many times you can figure out an unknown word by reading on to the end of the sentence or story. Think: "What word begins like the unknown word and makes sense in this sentence or story?"
13. For some story problems, you may be able to skim the story to find the answer. *Skim* here means *to quickly glance through the story until you come to the part that answers the question.*

Know When to Guess on a Multiple-Choice Test

14. After reading the question and each of the possible answers, some of the answers may look wrong to you. Eliminate those answers from your thinking. This should leave you with fewer answers to consider.

15. If you are uncertain which of the remaining answers is correct, take a guess at the one you *think* is correct. But if they all seem correct, don't guess–just go on to the next item.

Mark Your Answers Carefully

16. Your teacher may ask you to use a separate answer sheet for some tests. Some answer sheets are scored by a machine that cannot tell the difference between your answer and stray marks. No "doodling" on answer sheets because it could lower your score.

17. Find out which way the numbers go. Left to right? Top to bottom? (Not all answer sheets are alike.)

18. Keep your place on the answer sheet. Always check to make sure that your mark is put in the right place. If you are on question 19, you should make a mark by number 19 on the answer sheet.

19. If you need to change an answer, be sure to erase your mark cleanly.

Keep Track of Time

20. If you have a set amount of time to complete a test or section, look at the clock (or the number of minutes your teacher writes on the chalkboard) to find out about how much time you have left.

21. Don't waste time while you are working. Move quickly from one question to another. But don't work *too* fast—or you may make careless errors.

22. While you are working, check to see how much time you have left. If you are running out of time and have many questions left to do, you will need to work faster. If you have only a minute or two left, glance at the remaining problems to find the easiest ones. Answer those and skip the rest.

Look Over Your Answer Sheet

23. When you finish the test, look over your answer sheet. Be sure to erase any stray marks that remain.

24. If you have time, also look over your answers.

25. Above all, try not to disturb classmates who are still working.

Learning Goals

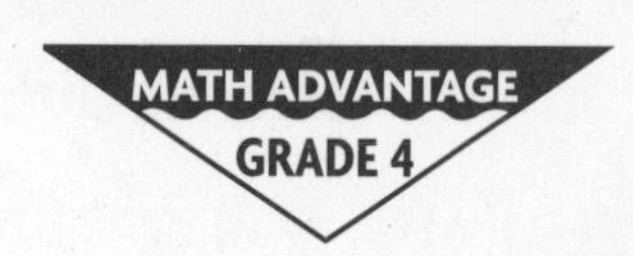

Student's Name ______________________ Teacher ______________

CHAPTER 1

Test Score:	Criteria 17/24 / Form A ____ / Form B ____	Needs More Work	Accom-plished
1-A.1 ☐ ☐	To use addition and sub-traction to make equations equal		
1-A.2 ☐ ☐	To find sums by adding three or more addends		
1-A.3 ☐ ☐	To use rounding to estimate sums and differences		
1-A.4 ☐ ☐	To add and subtract two-, three-, and four-digit numbers and money amounts with and without regrouping		

CHAPTER 2

Test Score:	Criteria 14/20 / Form A ____ / Form B ____	Needs More Work	Accom-plished
2-A.1 ☐ ☐	To use addition and subtraction to solve problems		
2-A.2 ☐ ☐	To subtract across zeros with multiple regroupings		
2-A.3 ☐ ☐	To use rounding to estimate sums and differences		
2-A.4 ☐ ☐	To choose the operation to solve problems and then find the sums or differences		

CHAPTER 3

Test Score:	Criteria 14/20 / Form A ____ / Form B ____	Needs More Work	Accom-plished
3-A.1 ☐ ☐	To model multiplication problems		
3-A.2 ☐ ☐	To use the Property of One, the Zero Property, and the Order Property to find products		
3-A.3 ☐ ☐	To multiply three factors		
3-A.4 ☐ ☐	To choose the operation to solve problems		

CHAPTER 4

Test Score:	Criteria 14/20 / Form A ____ / Form B ____	Needs More Work	Accom-plished
4-A.1 ☐ ☐	To decide whether a problem can be solved by using division		
4-A.2 ☐ ☐	To use inverse operations to solve problems		
4-A.3 ☐ ☐	To divide with remainders and make a table to solve problems		
4-A.4 ☐ ☐	To use a multiplication table to find quotients		
4-A.5 ☐ ☐	To divide with 1 and practice basic division facts		
4-A.6 ☐ ☐	To use the strategy *choose the operation* to solve problems		

Learning Goals

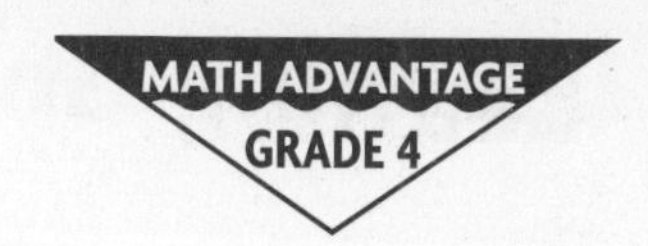

Student's Name ________________________ Teacher ____________

CHAPTER 5

Test Score:	Criteria 14/20 Form A ____ Form B ____	Needs More Work	Accom-plished
5-A.1 ☐ ☐	To learn different meanings and uses of numbers		
5-A.2 ☐ ☐	To model, read, and write numbers to 9,999		
5-A.3 ☐ ☐	To use benchmark numbers to show relationships among numbers		

CHAPTER 6

Test Score:	Criteria 14/20 Form A ____ Form B ____	Needs More Work	Accom-plished
6-A.1 ☐ ☐	To identify place value and represent numbers through 9,999 in standard, expanded, and written form		
6-A.2 ☐ ☐	To read and write numbers to millions		

CHAPTER 7

Test Score:	Criteria 14/20 Form A ____ Form B ____	Needs More Work	Accom-plished
7-A.1 ☐ ☐	To use a number line to compare numbers		
7-A.2 ☐ ☐	To use place value to compare numbers		
7-A.3 ☐ ☐	To use number lines and place value to order numbers		
7-A.4 ☐ ☐	To sort and compare sets of numbers, using a Venn diagram		

CHAPTER 8

Test Score:	Criteria 14/20 Form A ____ Form B ____	Needs More Work	Accom-plished
8-A.1 ☐ ☐	To tell time to the minute and second and give reasonable units of time for activities		
8-A.2 ☐ ☐	To tell the difference between *A.M.* and *P. M.*		
8-A.3 ☐ ☐	To read and use a schedule to solve problems		
8-A.3 ☐ ☐	To use a calendar to measure elapsed time		

Learning Goals

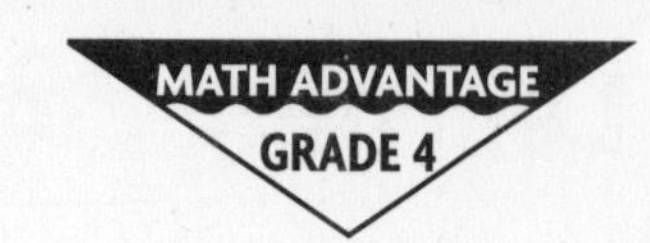

Student's Name ______________________ Teacher ______________

CHAPTER 9

Test Score:	Criteria 10/15 Form A ___ Form B ___	Needs More Work	Accom-plished
9-A.1 ☐ ☐	To organize data in tables and use the data to solve problems		
9-A.2 ☐ ☐	To evaluate surveys for usefulness in appropriate data		
9-A.3 ☐ ☐	To interpret data in bar graphs		

CHAPTER 10

Test Score:	Criteria 14/20 Form A ___ Form B ___	Needs More Work	Accom-plished
10-A.1 ☐ ☐	To read and interpret data on a line graph		
10-A.2 ☐ ☐	To read and interpret data on a line plot		
10-A.3 ☐ ☐	To read and interpret data on a stem-and-leaf plot		
10-A.4 ☐ ☐	To choose an appropriate graph to represent data		

CHAPTER 11

Test Score:	Criteria 12/18 Form A ___ Form B ___	Needs More Work	Accom-plished
11-A.1 ☐ ☐	To determine whether events are certain or impossible, likely or unlikely		
11-A.2 ☐ ☐	To determine the probability of an event		
11-A.3 ☐ ☐	To determine whether or not a game is fair		

CHAPTER 12

Test Score:	Criteria 14/20 Form A ___ Form B ___	Needs More Work	Accom-plished
12-A.1 ☐ ☐	To identify and distinguish between one-, two-, and three-dimensional figures and their properties		
12-A.2 ☐ ☐	To locate points on a coordinate grid and identify a plane figure made by connecting points		

Learning Goals

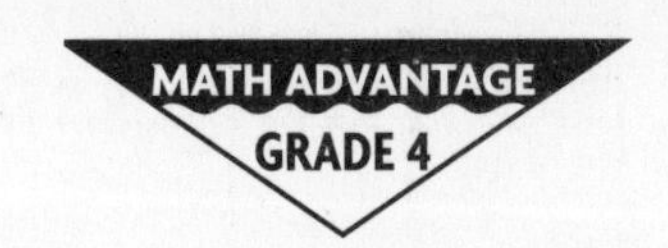

Student's Name ______________________ Teacher ______________

CHAPTER 13

Test Score:	Criteria 14/20 · Form A ____ · Form B ____	Needs More Work	Accom-plished
13-A.1 ☐ ☐	To identify points, planes, lines, line segments, angles and angle relationships		
13-A.2 ☐ ☐	To identify and classify polygons		
13-A.3 ☐ ☐	To identify and classify quadrilaterals		

CHAPTER 14

Test Score:	Criteria 14/20 · Form A ____ · Form B ____	Needs More Work	Accom-plished
14-A.1 ☐ ☐	To find the perimeter of a figure		
14-A.2 ☐ ☐	To find the area of regular and irregular figures		
14-A.3	To find the perimeter and area of the same figure and predict how changing the shape of the figure will change its area and perimeter		

CHAPTER 15

Test Score:	Criteria 12/18 · Form A ____ · Form B ____	Needs More Work	Accom-plished
15-A.1 ☐ ☐	To translate, reflect, and rotate figures and determine that the figure's size and shape do not change		
15-A.2 ☐ ☐	To identify and construct congruent figures		
15-A.3 ☐ ☐	To identify figures with point symmetry and figures with line symmetry		
15-A.4 ☐ ☐	To identify tessellations and properties of figures that tessellate		
15-A.5 ☐ ☐	To identify similar figures and solve problems by using the strategy *make a model*		

CHAPTER 16

Test Score:	Criteria 14/20 · Form A ____ · Form B ____	Needs More Work	Accom-plished
16-A.1	To multiply a one-digit number by 10, 100, and 1,000		
16-A.2	To multiply a one-digit number by a two- or three-digit number		
16-A.3	To multiply a whole number by amounts of money to solve problems		

Learning Goals

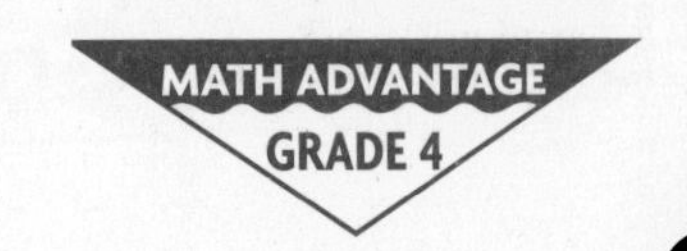

Student's Name ______________________ Teacher ______________

CHAPTER 17

Test Score:	Criteria Form A Form B 12/20 ____ ____	Needs More Work	Accom-plished
17-A.1 ☐ ☐	To use basic facts and patterns of tens to multiply two-digit numbers and solve problems		
17-A.2 ☐ ☐	To estimate products of two-digit numbers by rounding factors		
17-A.3 ☐ ☐	To multiply by two-digit numbers using partial products		
17-A.4 ☐ ☐	To solve problems by multiplying two-digit numbers		

CHAPTER 18

Test Score:	Criteria Form A Form B 13/20 ____ ____	Needs More Work	Accom-plished
18-A.1 ☐ ☐	To use basic facts to divide with remainders		
18-A.2 ☐ ☐	To divide a two-digit number by a one-digit number without a remainder		
18-A.3 ☐ ☐	To divide by two-digit numbers using partial products		
18-A.4 ☐ ☐	To solve problems by dividing two- and three-digit numbers		

CHAPTER 19

Test Score:	Criteria Form A Form B 20/24 ____ ____	Needs More Work	Accom-plished
19-A.1 ☐ ☐	To use basic facts and patterns of zero to estimate quotients		
19-A.2 ☐ ☐	To divide when a zero is in the dividend or in the quotient		
19-A.3 ☐ ☐	To divide amounts of money		
19-A.4 ☐ ☐	To interpret the remainder when solving division problems		
19-A.5 ☐ ☐	To choose the operation when solving problems		

CHAPTER 20

Test Score:	Criteria Form A Form B 16/24 ____ ____	Needs More Work	Accom-plished
20-A.1 ☐ ☐	To identify the fractional part of a whole or group		
20-A.2 ☐ ☐	To compare and order fractions		
20-A.3 ☐ ☐	To identify, read, and write mixed numbers and change fractions greater than 1 to mixed numbers		

Learning Goals

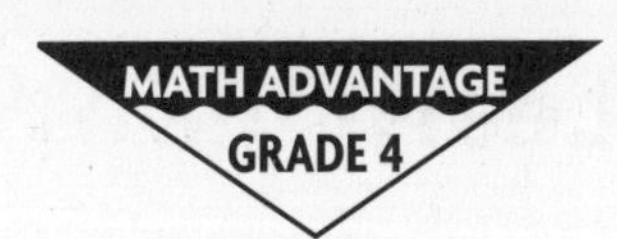

Student's Name ______________________ Teacher ______________________

CHAPTER 21

Test Score:	Criteria 16/24 Form A ____ Form B ____	Needs More Work	Accomplished
21-A.1 ☐ ☐	To solve problems by adding fractions and mixed numbers with like denominators		
21-A.2 ☐ ☐	To solve problems by subtracting fractions and mixed numbers with like denominators		

CHAPTER 22

Test Score:	Criteria 15/22 Form A ____ Form B ____	Needs More Work	Accomplished
22-A.1 ☐ ☐	To relate fractions and decimals		
22-A.2 ☐ ☐	To identify, read, and write decimals to hundredths		
22-A.3 ☐ ☐	To identify, read, and write mixed decimals		
22-A.4 ☐ ☐	To compare and order decimals and mixed decimals		

CHAPTER 23

Test Score:	Criteria 16/24 Form A ____ Form B ____	Needs More Work	Accomplished
23-A.1 ☐ ☐	To solve problems by adding and subtracting decimals		
23-A.2 ☐ ☐	To estimate decimal sums and differences by rounding		

CHAPTER 24

Test Score:	Criteria 15/22 Form A ____ Form B ____	Needs More Work	Accomplished
24-A.1 ☐ ☐	To measure length in linear units to the nearest unit or fraction of a unit		
24-A.2 ☐ ☐	To use multiplication and division to change units of measure		
24-A.3 ☐ ☐	To use diagrams to solve problems		
24-A.4 ☐ ☐	To use appropriate units to measure weight		

Learning Goals

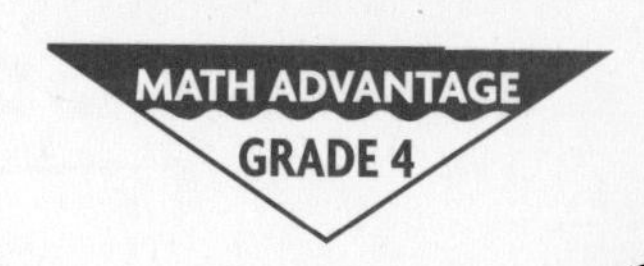

Student's Name ______________________ Teacher ______________

CHAPTER 25

Test Score:	Criteria 18/24 / Form A ____ / Form B ____	Needs More Work	Accomplished
25-A.1 ☐ ☐	To use multiplication to change units of measure		
25-A.2 ☐ ☐	To use the strategy *solve a simpler problem* to solve problems		
25-A.3 ☐ ☐	To use appropriate units to measure capacity		

CHAPTER 26

Test Score:	Criteria 14/20 / Form A ____ / Form B ____	Needs More Work	Accomplished
26-A.1 ☐ ☐	To express time by using fractions of an hour		
26-A.2 ☐ ☐	To compare customary and metric linear units of measure		
26-A.3 ☐ ☐	To *write a number sentence* to solve problems		
26-A.4 ☐ ☐	To read Fahrenheit and Celsius thermometers and compare the difference in degrees of two temperatures		

CHAPTER 27

Test Score:	Criteria 17/24 / Form A ____ / Form B ____	Needs More Work	Accomplished
27-A.1 ☐ ☐	To use multiples of 10 and a pattern of zeros to estimate quotients		
27-A.2 ☐ ☐	To divide by multiples of 10		
27-A.3 ☐ ☐	To divide three-digit numbers by two-digit divisors		
27-A.4 ☐ ☐	To correct the estimated quotient if it is too high or too low		
27-A.5 ☐ ☐	To *write a number sentence* to solve problems		

CHAPTER 28

Test Score:	Criteria 12/18 / Form A ____ / Form B ____	Needs More Work	Accomplished
28-A.1 ☐ ☐	To use fractions to identify data represented in a circle graph		
28-A.2 ☐ ☐	To use decimals to identify data represented in a circle graph		
28-A.3 ☐ ☐	To choose a graph that represents the given data